Little Penguin

Julie Murray

Abdo Kids Junior
is an Imprint of Abdo Kids
abdobooks.com

abdobooks.com

Published by Abdo Kids, a division of ABDO, P.O. Box 398166, Minneapolis, Minnesota 55439. Copyright © 2020 by Abdo Consulting Group, Inc. International copyrights reserved in all countries. No part of this book may be reproduced in any form without written permission from the publisher. Abdo Kids Junior™ is a trademark and logo of Abdo Kids.

Printed in the United States of America, North Mankato, Minnesota.
102019
012020

THIS BOOK CONTAINS RECYCLED MATERIALS

Photo Credits: iStock, Minden Pictures, Science Source, Shutterstock

Production Contributors: Teddy Borth, Jennie Forsberg, Grace Hansen

Design Contributors: Christina Doffing, Candice Keimig, Dorothy Toth

Library of Congress Control Number: 2019941219

Publisher's Cataloging-in-Publication Data

Names: Murray, Julie, author.
Title: Little penguin / by Julie Murray
Description: Minneapolis, Minnesota : Abdo Kids, 2020 | Series: Mini animals | Includes online resources and index.
Identifiers: ISBN 9781532188817 (lib. bdg.) | ISBN 9781644943038 (pbk.) | ISBN 9781532189302 (ebook) | ISBN 9781098200282 (Read-to-Me ebook)
Subjects: LCSH: Little penguin--Juvenile literature. | Animals--Australia--Juvenile literature. | Penguins--Juvenile literature. | Southern blue penguin--Juvenile literature. | Size and shape--Juvenile literature.
Classification: DDC 598.441--dc23

Table of Contents

Little Penguin 4

Let's Compare! 22

Glossary. 23

Index 24

Abdo Kids Code. 24

Little Penguin

It is also called a blue penguin.

It is the smallest penguin.

It lives in Australia. It also lives in New Zealand.

It is 12 inches (30 cm) tall.

It weighs 3 pounds (1.4 kg).

It has blue feathers.

Its belly is white.

It has two **flippers**. Its feet are **webbed**. These help it swim.

It swims most of the day.

It rests at night.

15

It makes a barking noise.

This helps it find others.

It **molts** each year. It grows new feathers.

It dives for food. It eats fish and squid. It also eats krill.

Let's Compare!

little penguin

emperor penguin

Height: 13 inches (33 cm)
Weight: 3.3 pounds (1.5 kg)

Height: 48 inches (122 cm)
Weight: 49 to 99 pounds
(22 to 45 kg)

Glossary

molt
to give off feathers from the body and grow a new covering.

flipper
a wide, flat limb on some animals that is used for swimming.

webbed
having toes connected by skin.

Index

Australia 6

color 10

communication 16

feathers 10, 18

feet 12

flippers 12

food 20

habits 14

New Zealand 6

size 4, 8

swimming 12, 14

Visit abdokids.com to access crafts, games, videos, and more!

Use Abdo Kids code MLK8817 or scan this QR code!